SUPERSTAR

SUPER
The Supernova

A Voyage into Space Book   •   **Franklyn M. Branley**

illustrations by True Kelley   •   **Thomas Y. Crowell   New York**

STAR

of 1987

My thanks to so many;
especially Kathleen Zoehfeld for her skillful editing,
and Stephen P. Maran for his expert suggestions.
—F. M.B.

SUPERSTAR: *The Supernova of 1987*
Copyright © 1990 by Franklyn M. Branley
Illustrations copyright © 1990 by True Kelley
Printed in the U.S.A. All rights reserved.
Typography by Elynn Cohen
1 2 3 4 5 6 7 8 9 10
First Edition

Library of Congress Cataloging-in-Publication Data
Branley, Franklyn Mansfield, date
  Superstar : the Supernova of 1987 / by Franklyn M. Branley ;
illustrated by True Kelley.
     p.   cm. — (A Voyage into space book)
  Includes bibliographical references.
  Summary: Describes the nature and origin of supernovas, how they
provide information on the formation of stars and planets, and what
was learned from study of Supernova 1987A.
  ISBN 0-690-04839-4 : $. — ISBN 0-690-04841-6 (lib. bdg.) :
$
  1. Supernova 1987A—Juvenile literature.   2. Supernovae—Juvenile
literature.   [1. Supernova 1987A   2. Supernovae.]   I. Kelley,
True, ill.   II. Title.   III. Series.
QB843.S95B73   1990   89-71164
523.8′446—dc20   CIP
                                                        AC

Photo Credits:
© Ian Shelton (p. vi, top and bottom; color insert 1, top); © Kathy White
(color insert 1, bottom); © Royal Observatory, Edinburgh/Anglo-Austra-
lian Telescope Board, Photographs by David Malin from original plates
taken with the UK Schmidt telescope, New South Wales, Australia (color
insert 3, top); Smithsonian Institution Photo No. 82-4770 (p. 24); Yerkes
Observatory (p. 25); Lick Observatory, University of California (p. 29, top);
Joe Stancampiano and Karl Luttrell, © National Geographic Society (color
insert 3, bottom); © European Southern Observatory (color insert 4;
p. 51).

CONTENTS

(Color photo insert begins after page 26)

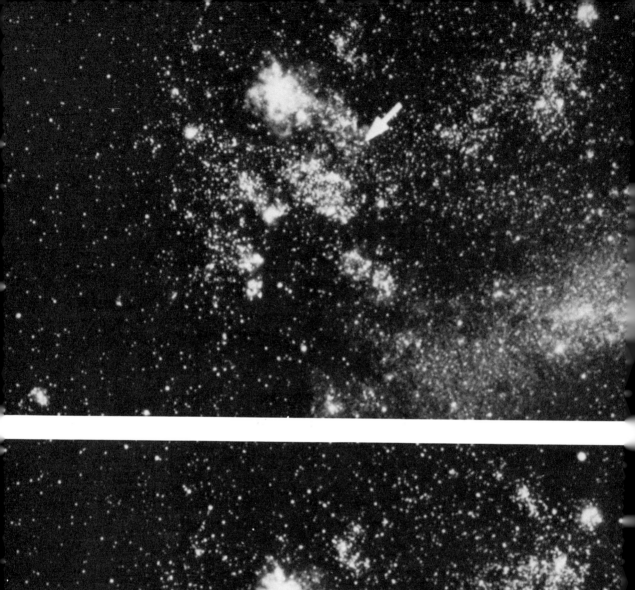
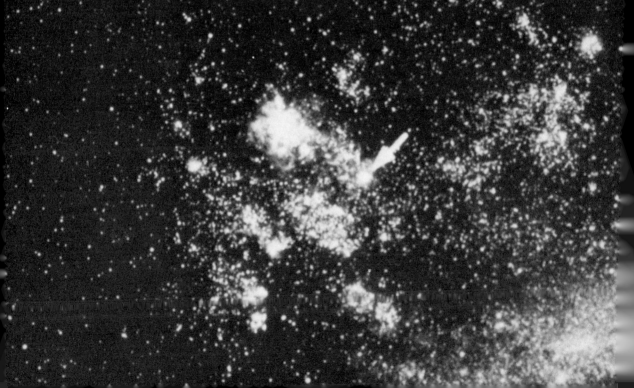

1

DISCOVERY

On February 24, 1987, Ian Shelton, a young Cana-
dian working at Las Campanas Observatory in northern Chile, aimed a telescope-camera at the Large Magellanic Cloud (LMC). This is a nearby galaxy, some
170,000 light-years away. After making a three-hour
exposure, Shelton decided to develop the film right
away. The usual procedure was to develop films later

*On the opposite page below is Ian Shelton's "discovery" photograph—the
one in which the supernova was first detected. Supernova 1987A is the bright
star to the right of center. It is in the Large Magellanic Cloud, a hazy galaxy
that is some 170,000 light-years away. The bright patch to the left of the
supernova is a gaseous formation, or nebula.*

   *The picture above, also taken by Ian Shelton, shows the LMC before the
supernova appeared. The arrow points to the star that exploded—the progenitor,
or producer, of the supernova.*

in the day.

He had photographed the LMC on other occasions, so he knew how it should appear. But this picture was different. There was a bright star in the Cloud, brighter than the galaxy itself. Shelton was not happy. He thought there must have been a flaw in the film, or perhaps he had made a mistake when developing it. He knew there was no star that bright in the LMC.

He stepped outside and looked at the Cloud. The bright star was there. He could see it clearly. His photograph was correct. A star in the LMC had suddenly become very bright, brighter than a hundred million average-sized stars like the Sun. The star was a supernova, the brightest one seen in the past 383 years. And Ian Shelton, a twenty-nine-year-old worker from the University of Toronto, had discovered it.

Nova means "new." Astronomers use nova as shorthand for *stella nova,* or "new star." Novas are stars that appear where earlier no star had been visible. The star had *been* there but had been too dim for us to see. So although novas are not really new, they are new to a star observer. They are stars that suddenly become very bright, giving off much more energy than they had done previously. Should the amount of energy be much, much greater, the star is called a supernova.

Supernovas are not that unusual, for astronomers have identified several hundred of them. Most are in far-off galaxies, and although they are very bright actually, they appear dim because of their great distance from us. Occasionally, perhaps three or four times in a thousand years, a supernova is close enough to appear very bright—in rare cases, so bright it can be seen in daytime.

The new star in the LMC was named SN 1987A: SN for supernova, and 1987A because it was the first supernova seen in 1987.

2

SUPERSTARS ARE
STAR BUILDERS

Whether they are a few thousand light-years away or millions, supernovas are of tremendous interest. Many of the elements in the universe are created inside supernovas. Stars are made of gases—mostly hydrogen and helium, which were made when the universe formed and which are the two most common elements in the universe.

On occasion, great amounts of hydrogen pack together, forming a huge clump. As the gases pack closer together, the temperature at the center of the clump rises. The temperature goes up and up, reaching ten million degrees. At that temperature, the cores of hydrogen atoms, or hydrogen nuclei, begin to join

A CROSS SECTION OF THE SUN

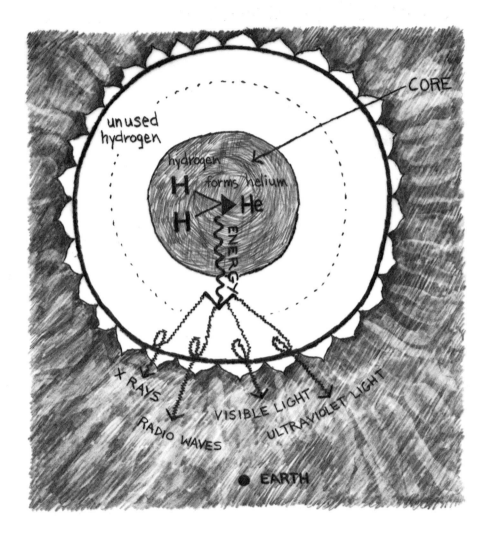

The Sun is a hydrogen-fusing star. Hydrogen nuclei in the center of the Sun fuse to produce helium. In the process, energy is released.

together, or fuse, producing helium nuclei, in what is called a nuclear fusion reaction. As this occurs, energy is released as radio waves, heat, X rays, gamma rays, ultraviolet light, and visible light. The clump of gases has become a star.

The Sun is a hydrogen-fusing star. In the millennia ahead it will use up the hydrogen in its core. Then it will become a helium-fusing star. The helium nuclei

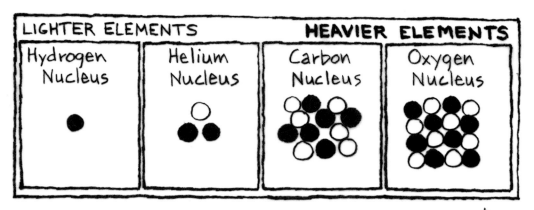

Hydrogen is the lightest of all the elements. It is element number, or atomic number, 1. It has one proton in its core, or nucleus. Helium is atomic number 2. It has two protons in its nucleus. The higher the atomic number, the more protons the element has in each of its atomic nuclei. And the more protons an element has in its nuclei, the heavier it is.

will fuse, producing the heavier nuclei of carbon and oxygen. At the same time, the Sun will swell to a huge size. It will become a red giant star. The red giant Sun may be large enough to reach beyond the orbit of Earth or even of Mars. Eventually, though, it will expel its outer layers. All that will be left will be a tiny white dwarf star—a sphere about the size of Earth. In the Sun, the fusion reactions will end with the production of carbon and oxygen from helium.

In more massive stars, the central temperatures and densities can get much higher than those inside the Sun. When that happens, oxygen and carbon also fuse, producing nuclei of even heavier elements such as sulfur, silicon, nickel, and iron. Once the central part, or core, of a massive star becomes iron, however, its energy supply stops. It takes more energy to fuse iron than is given off when iron forms. So iron is not a nuclear energy source. Elements heavier than iron, which is element number, or atomic number, 26, cannot be produced in such stars.

In the universe, there are over a hundred elements. Eighty or so have atomic numbers greater than 26—elements such as zinc, silver, tin, and gold. Many of the elements with atomic numbers higher than iron are found in the Sun and in other stars. But if the heavy elements cannot be made in those stars, where could they have come from?

Theory says they must come from a location where there is a lot of raw material—that is, elements up to number 26—and where temperature is measured in billions of degrees—a thousand times greater than the temperature of the Sun. One place where these conditions exist is in a supernova.

Until 1987, such an explanation was only a theory; no one had ever seen such things happen. That's because all supernovas observed in modern days have been far away—so far away that they were dim and could not be studied in detail even with advanced instruments.

Down through history, there have been bright supernovas—in A.D. 1006, 1054, 1572, and 1604—but there were no accurate instruments for observing them. Astronomers didn't even have telescopes, for the telescope was not invented until 1609. Observers of these early supernovas could study them only with their unaided eyes. Therefore, little was learned about them other than that they changed in brightness.

Supernova 1987A gave astronomers the first opportunity to study a bright and relatively nearby supernova with cameras, light meters, spectroscopes for analyzing light, and other modern instruments—many of them aboard satellites. Astronomers could watch a supernova from its beginning and follow changes as they occurred. It became possible for astronomers to prove that at least some of the elements heavier than iron are created in the superhot interiors of supernovas during cataclysmic explosions. Supernovas are the cradles of the elements.

Even before SN 1987A, many scientists theorized that the Sun, the planets, and our entire solar system formed out of a cloud of cosmic gases and dust. Most of the cloud was hydrogen, but mixed into it were heavy elements produced by ancient supernova blasts. At some point in time a shock wave produced by the explosion of one such supernova passed through the cloud. The shock wave compressed great masses of the dust and gases, making clumps that grew to become the Sun and the planets.

According to the theory, remnants of supernovas, including the elements beyond iron, become the substance of new stars. All the elements up to iron as well as the newly made elements are added to the basic hydrogen and helium of the universe. The new stars formed of these elements are second-, third-, or even fourth-generation stars—they are made of "used" material, material that at one time was in other stars. The Sun is an example—it is a "secondhand" star. Planets that may form around a star also contain atoms made during the supernova explosions. People on those planets are made of the same materials. So each of us is made of products that were formed during supernova explosions. We are all, in a sense, children of the stars.

Supernovas are the builders of stars and planets, and of people—that had been the theory. By observing SN 1987A, astronomers have already been able to prove much of the theory.

One of the elements identified in the energy given off by SN 1987A is cobalt 56. This is a form of cobalt that breaks down, or decays, to become iron. In the decay process, gamma rays are released. The intensity of the gamma radiation coming from SN 1987A indicates the mass of the cobalt produced is equal to 80 times the mass of Jupiter. Astronomers know that the cobalt is new; it was created when the explosion of SN 1987A occurred. They know this because the half-life decay period of cobalt 56 is only 77 days. In that time half of the cobalt 56 would become iron. In only a year or so, all of the cobalt 56 would disappear. Other elements, such as nickel, silicon, and neon have also been identified. They all serve to strengthen the theory about the creation of heavy elements in supernova explosions and to prove that many elements, including those created during the lifetime of a star, are ejected into space by the explosions.

3

WHY DO
STARS EXPLODE?

Stars are massive formations of gases. In that respect, they are all similar. In other ways, stars differ from one another. They may be large or small, hot or cool, bright or dim, young or old; and they may be white, blue, yellow, orange, or red.

All stars, regardless of size or mass, color or temperature, have life histories. Stars get older and go out of existence. At the same time, other stars are being created. Whether a star has a short lifetime or a long

A star has strong gravitation, which tends to pull its gases inward, causing the star to collapse. At the same time, high temperatures create outward pressure, causing the star to expand.

The two forces offset each other. Thus a star such as the Sun is in balance; it is not collapsing, nor is it expanding.

one depends on the mass of the star—the amount of material it contains.

The mass of the Sun is 300,000 times Earth's mass. Even so, the Sun is only a medium star. It has a lifetime of some 10 billion years. Presently the Sun is about 5 billion years old; it is a middle-aged star.

Stars less massive than the Sun are also cooler. They use up nuclear fuel slowly, and therefore they have lifetimes much longer than the Sun's. Stars more massive than the Sun are also hotter. They use material rapidly, and therefore have shorter lifetimes.

Stars that shine steadily, such as the Sun, are in balance. Pressure generated by the high inside temperature pushes outward. At the same time, gravity pulls inward. The force outward is balanced by the force inward—the star is in equilibrium. Should the force inward become greater, the star collapses and heats rapidly to very high temperatures. Should the force outward become greater, the star expands, or it may explode. Large amounts of it may be blown into space. Such imbalances occur often—there are always many collapsing stars as well as many that are expanding or exploding. Stars are always going out of existence in one way or another.

Type I Supernovas

Modern astronomers have studied many exploding stars. No two behave in exactly the same way, for each one is different. However, they are similar enough to be classified into two large groups—Type I and Type II. A Type I event occurs when two stars are involved. While stars may appear to be single to the unaided eye, through a telescope many are revealed to be binary stars—two stars that go around each other. One of the stars may be large and bright and easy to see. The other may be dim and small—an invisible white dwarf star.

The white dwarf star has high mass and high density (material in it is packed together tightly) and very strong gravity. If a white dwarf's companion star has lower density, where gases are not held together tightly, the strong gravity of the white dwarf will pull gases away from the larger, brighter companion. The gases stream into and around the white dwarf; mass is taken from the larger star and added to the smaller one. The temperature of the white dwarf is not high enough to exert much outward pressure; therefore,

TYPE I SUPERNOVA

gravitation packs the atoms of the newly added gases even more tightly. Mass increases until it becomes too much for the white dwarf to maintain equilibrium. Immense gravity causes the star to collapse. As this happens, temperature soars to several billion degrees. The star blows apart. Just about all the matter in it is ejected into space. Outer layers of gases from the companion star may also be blown away. A tremendous gas cloud is produced. The cloud emits radio waves and X rays in great amounts, which diminish rapidly as the cloud expands. This kind of explosion, one that involves two stars, is called a Type I supernova. For a day or two some astronomers believed SN 1987A was this kind of supernova. After more careful study, all agreed that it was not. The explosion appeared to be more like that of a Type II event.

A Type I supernova occurs in a double-star system. One star is very dense and small and has strong gravity. It pulls gases from a gigantic companion star.

As the small white dwarf star accumulates gases, the gravitation of the dwarf increases rapidly. The gases collapse, temperature soars, and a tremendous explosion occurs. It blows the star apart, releasing fantastic amounts of energy. All that remains is a huge, gaseous, energy-releasing cloud.

Type II Supernovas

A Type II supernova is produced by a star that has no companion star affecting it. The mass of such a star must be at least eight times greater than the Sun's mass. The star uses hydrogen in its interior rapidly, perhaps taking only a few million years or even less to use nearly all of its core hydrogen. Once this has been done, the star begins to contract. As it packs together, temperature can skyrocket to 180 million degrees or more. The more massive the star, the hotter it gets at the center. (Temperature at the center of the Sun is 15 million degrees.) Helium nuclei fuse to become carbon and oxygen nuclei. In the fusion process, more energy is released, causing the star to expand. But once the helium fuel is consumed, fusion reactions come to a halt, and the temperature drops

A Type II supernova occurs in a star much more massive than the Sun. It uses nuclear fuel rapidly. When fuel for nuclear reactions is consumed, the star collapses. The tremendous inward collapse causes a rebound—a shock wave that explodes outward, blowing away most of the star.

During the collapse, electrons combine with protons, producing neutrons. The remnants of the star are a huge, expanding cloud of gases and a small, dense neutron star.

TYPE II SUPERNOVA

1 SUPERGIANT IN BALANCE

2 IT COLLAPSES INTO A NEUTRON STAR

3 A SHOCKWAVE REBOUNDS

4

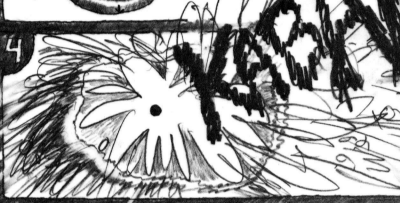

KABOOM

5 SHOCKWAVE

NEUTRON STAR   expanding gas cloud

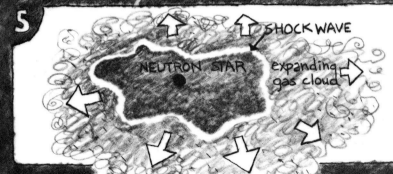

fairly quickly. Pressure outward drops, and the star collapses. As the material packs together, the temperature goes up even higher than it had been before. It gets hot enough for the new carbon and oxygen nuclei to fuse into heavier elements. This process of expanding and contracting/heating and cooling continues, each step taking less and less time to be completed. Finally, silicon in the star fuses to become nickel, cobalt, and iron.

The star has an iron core. Outside the core are layers of other nuclei—silicon, magnesium, neon, oxygen, carbon, helium, and hydrogen. Each layer is formed of leftover fuel from previous fusion reactions. The outermost layer is hydrogen, the lightest material. It becomes the atmosphere or visible "surface" of the star.

The iron does not fuse; thus, fusion energy is no longer generated. Outward pressure drops. But the star is massive; it has strong gravity, which in only a few moments causes the star to collapse. The core, which is about the size of Earth, collapses further to only about 10 miles in diameter. The inward collapse causes a rebound: The core expands to about 15 miles, and a shock wave moves outward from the core. Scientists believe that as the shock wave collides with the outer layers of the star, it causes a mammoth explosion, throwing the outer layers into space. Very high temperatures caused by the collapse and the shock wave cause nuclei to fuse and to form many elements, including those heavier than iron. Perhaps 80 percent of the star is ejected. The other 20 percent packs together tightly, producing a degenerate, or highly compressed, star that may become a neutron star.

These theories about supernovas, the creation of elements, and the creation of neutron stars grew rapidly in the 1960s and 1970s—much more rapidly than did observations to support the theories. Most supernovas cannot be observed in detail, because they are far away in distant galaxies or they are shielded from view by dense clouds in our own galaxy. Until 1987, only four supernovas had been bright enough for easy and detailed observation. As mentioned earlier, all occurred before telescopes and other instruments to study stars and planets were invented; therefore, little was learned about these dramatic stars.

4

EARLY GUEST STARS

Through the centuries, the Chinese have been careful sky watchers. They were familiar with the stars and knew the sky well enough to be keenly aware of changes in it. When a new star appeared, as happened occasionally, the visitor was called a guest star. While several lesser ones have been recorded, we shall discuss the four of most interest to the Chinese, and to observers in other parts of the world as well.

On April 30, 1006, a bright star was seen in the constellation Lupus, the Wolf. It is classified as a southern formation. Even though Lupus is in the southern skies, the bright new star was reported by observers in Egypt, Japan, China, and southern Europe. For all of them, it must have been very close to the southern horizon. This was the brightest of all supernovas, more luminous than the planet Venus at its brightest. According to one report, the brightness was equal to that of a quarter moon. A remnant of the supernova has recently been detected as a gas cloud, or nebula, in Lupus that gives off radio waves, X rays, and also dim visible light.

Tycho Brahe, a sixteenth-century Danish astronomer, was the most skillful observer of his time. In November of 1572 he saw a brilliant star appear in the constellation Cassiopeia, the Queen. Tycho kept careful records of changes in brightness of the star. From those records, astronomers of today believe he saw a Type I supernova, one occurring in a binary star system. No light can be seen today in Cassiopeia, but a remnant of that supernova—a gas cloud thrown off in its explosion—has been detected by the radio waves and X rays it gives off.

In 1604 Johannes Kepler, a German astronomer who had been one of Tycho's assistants, sighted a bright star in the constellation Ophiuchus, the Serpent Bearer. While the star was not as bright as Venus, it remained visible for a year. Indications are that it,

Tycho Brahe, a Danish astronomer (above), saw a supernova in 1572. He watched it and kept careful records of changes in brightness.

In 1604 Johannes Kepler, the German astronomer who had been an assistant to Tycho Brahe, discovered the last of the bright supernovas before SN 1987A.

24

CARL
NORDLING
del:

too, was a Type I event. In the Ophiuchus region there is now a gas cloud that produces radio waves and X rays. Judging by the location, it is very likely a remnant of the explosion seen by Kepler.

Of all the supernovas that we know about, by far the most interesting, and the one whose remnants have been most carefully studied, is the one that occurred in 1054.

The Crab Nebula

The guest star was reported by the Chinese on July 4, 1054. They saw a brilliant star appear near the tip of the lower horn of Taurus, the Bull. This was in the days when such sightings caused great concern, for people believed that all sky events had strong effects upon them. Sky events were most often considered signs of calamities such as floods, famines, earthquakes. In rare cases, people believed they foretold happy events such as good harvests or victories in battle.

The supernova of 1054 reached a brightness, or

The dome of Las Campanas Observatory sits high atop a 7,800-foot peak in the Andes Mountains in northern Chile. The building with the barnlike roof on the right holds the telescope-camera Ian Shelton used when he took the first photograph of SN 1987A.

In the darkroom where he developed the first photograph of SN 1987A, Ian Shelton compares his "before" and "after" shots.

The photograph above was taken shortly after the explosion of SN 1987A. The photograph below shows the same area of the Large Magellanic Cloud before the explosion. The arrow points to SN 1987A's progenitor, a blue supergiant star called Sanduleak −69° 202.

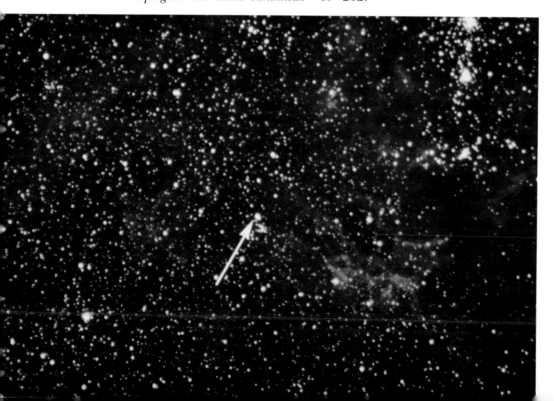

This picture, from a star atlas made in 1603, shows the location of Tycho Brahe's supernova of 1572 in the constellation Cassiopeia. The elaborate circle on the left represents the supernova.

This neutrino detector is a tank of pure water located far underground in an abandoned salt mine. Hours before SN 1987A was seen, the tank detected a shower of neutrinos, just as would be expected if there was a nearby supernova.

This color-enhanced photograph shows SN 1987A and the regions of the Large Magellanic Cloud surrounding it. The supernova is just to the right of center. The spikes of light that radiate from the supernova are called "diffraction effects." Diffraction effects are caused by the telescope when an extremely bright point of light is photographed through it. The large gas cloud above and to the left of center is called the "Tarantula Nebula" because of its spiderlike shape.

No doubt there are many supernovas now exploding throughout the universe. Energy from them may reach us tomorrow, or it may take thousands of years for it to get here.

magnitude, of −5. The brightness of stars is measured in magnitudes after a system that was invented by Hipparchus, a Greek astronomer who lived in the second century B.C. He could discern six steps in brightness. The most bright were the most important, Hipparchus believed, and so they were of first magnitude. Those somewhat less bright were second magnitude, then third magnitude, and so on, to the dimmest stars, which were sixth magnitude. The higher the number, the dimmer the star.

Hipparchus could use only his eyes to measure brightness. Now we have instruments that can measure more precisely and that can see very dim objects. We know that stars can be brighter than first magnitude. For example, the magnitude of the Sun is −26. Many objects are also much dimmer than sixth magnitude. Powerful telescopes with cameras can "see" objects of magnitude 26. The Chinese guest star of 1054 with a magnitude of −5 was very bright indeed—so bright it could be seen in daytime.

When we look at the Taurus region now, we cannot see the star. However, with telescopes we can see a tremendous gas cloud where the star appeared. It is called the Crab nebula—a name that seemed appropriate to the formation in 1844 when William Parsons, an Irish astronomer, saw it through his telescope. The gases are expanding—they are moving away from the central region at a speed of some 800 miles a second. They have been expanding since 1054 and have become a cloud that is seven light-years across. (It takes light seven years to go from one edge to the other, traveling 186,000 miles a second.)

Pulsar in the Crab Nebula

The Crab nebula produces light, which means it gives off energy. Anything that releases energy year after year must be resupplied in some way, just as a light bulb needs a constant supply of electricity. Astronomers could not explain how the Crab kept shining for a thousand years—that is, not until 1968.

A telescope is needed to see the Crab nebula. However, you can find its location at the tip of the lower horn of Taurus, the Bull. Taurus lies just above and to the right of Orion.

The nebula is a cloud of expanding gases produced in A.D. 1054 when a star exploded. A rotating neutron star, or pulsar, which is a remnant of that exploded star, is located near the center of the nebula.

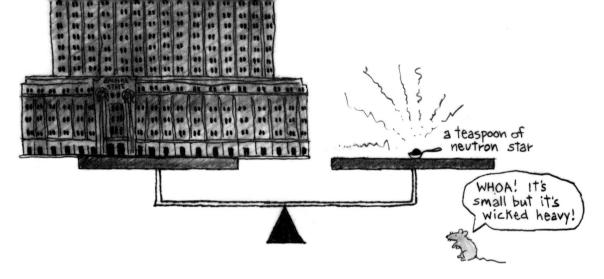

In neutron stars the particles of matter pack more and more tightly together. Density reaches incredible levels—just a teaspoon of neutron star matter could balance an object as heavy as the Empire State Building.

In that year, astronomers noticed that when they aimed their radio telescopes at the Crab, a rapid pulse of radio waves was received. The pulsations occurred 30 times a second. Further study found a star at the center of the Crab that apparently was the source of the energy. It was discovered that the light of the star also seems to pulse on and off 30 times a second. But how could a star change in brightness so rapidly?

It could if it was a neutron star. When a Type II supernova occurs, interior pressure and temperature become so high that electrons and protons are forced to combine. When an electron and a proton combine, a neutron is produced. Every electron carries a negative electric charge, and every proton has a positive charge. When they combine to form neutrons, the positive and negative charges cancel one another out. Because neutrons have no electric charge, they have no force to repel one another. They pack together closely, raising the density of the star very high. A star that may have been millions of miles across becomes one with a diameter of about 15 miles or less. Density becomes 100,000 tons per cubic millimeter—billions of tons per cubic inch.

Neutron stars spin rapidly—30 times a second in the case of the Crab. Careful study reveals that the Crab star is slowing down. As the star slows down, energy of rotation is released, and that energy appears as light, radio waves, and X rays. The energy is more than enough to keep the Crab nebula shining through the centuries.

A rotating neutron star is also called a pulsar. A pulsar has a strong electromagnetic field. The field is the source of a narrow beam of radio waves. As a pulsar rotates, its beam is turned alternately toward us and away from us. Each time the beam is directed toward us, astronomers can observe a flash, or pulse, of radio waves. As you can see from the diagram on p. 33, the effect is like that of a lighthouse. In a lighthouse, a big lamp turns, aiming its beam of light toward a particular spot once every few seconds. In the case of the Crab pulsar, the radio-wave beam pulses at the amazingly fast rate of 30 times a second. That's because the star turns 30 times a second.

Supernovas throw clouds of materials into space. Since the dawn of the universe, there have been millions of supernovas. Many of their expanding gas clouds have disappeared over time. But we should still be able to find a lot of gas clouds in the universe. And we have. Also, if the stellar remnant of a Type II supernova is a neutron star or a pulsar, there should be many neutron stars in the universe.

There probably are; however, only a few hundred have been observed. Others may be too far away to be seen. Still others may be almost impossible to detect because they have run down. They have lost their energy after millions of years. Another possibility is

---

Energy from a rotating neutron star, or pulsar, pulsates rapidly—30 or more times a second. One theory says that each time the pole of a star's magnetic field points toward Earth, we receive a burst of energy in the form of radio waves.

In Figure A, the pole of the star's magnetic field is toward Earth—a pulse is received. In Figure B, the star has made a one-half rotation; the pole is away from Earth—no pulse is received.

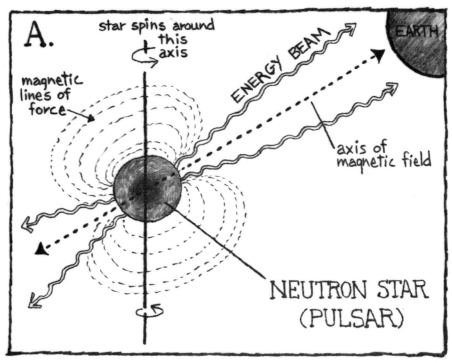

A.

star spins around this axis

magnetic lines of force

ENERGY BEAM

EARTH

axis of magnetic field

NEUTRON STAR (PULSAR)

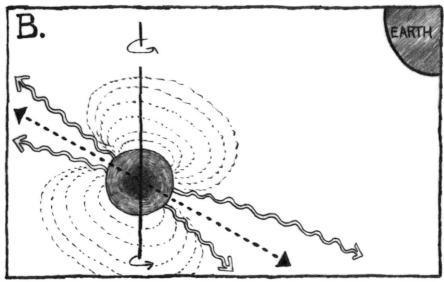

B.

EARTH

that our angle of viewing makes it impossible for us to pick up the pulses.

In 1989, two years after the sighting of SN 1987A, a few astronomers saw what they believe may be signs that a pulsar is shining through the expanding cloud of gas from the explosion. If they are right, then this will be the first time it has been possible to watch the development of a pulsar. The pulsar seems to be spinning at an unbelievable rate—making a turn in $\frac{1}{2,000}$ second. This means it is spinning at a speed of 200 million miles an hour. You'd expect this would be enough for the pulsar to fly apart. But it has tremendous density—about 500,000 Earth-masses compressed into a globe no larger in diameter than New York or Paris. Its density may give it enough gravity to hold the formation together in spite of its rapid rotation. However, the pulsar—if, indeed, that's what it is—seems to have a nearby companion. This may be a mass that has been thrown out from the main formation. We'll need more information before astronomers can explain it.

5

PHOTONS, NEUTRINOS, AND COSMIC RAYS

Supernova 1987A was a Type II event. That means hydrogen should be detected in the outer region; and it has been found. The International Ultraviolet Explorer, a satellite probe that is sensitive to short-wave energy, picked up signs of a huge shell of material, largely hydrogen, moving outward from SN 1987A, just as expected from a Type II supernova. The speed is about 9,300 miles a second.

Also, a shower of billions of neutrinos should have been detected before the light arrived. According to theory, nuclear fusion reactions in the cores of stars produce two kinds of particles that travel at the speed of light. One is called the photon, a unit of energy

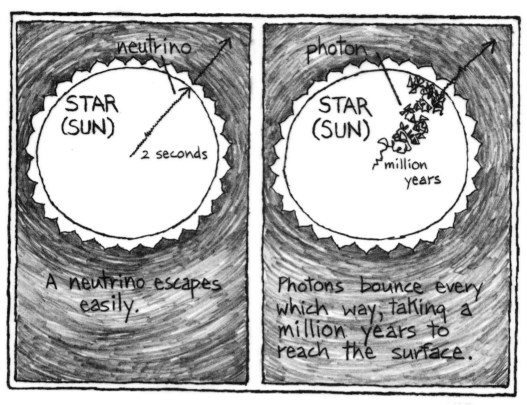

Neutrinos can go through the Sun, through Earth, and through you and me. Billions of them were released in the explosion of SN 1987A.

Photons cannot move freely through the Sun or through any other star. They bounce around inside the star and are absorbed and released many times, taking a million years to move through and out of the gases in a star.

such as light or X rays. The other is called the neutrino.

Photons are given off at the core of a star. They travel outward for a moment and are absorbed almost immediately. They are then released again, move for a moment, and are absorbed once more. Because of these absorption processes, it takes a million years or more for a photon to move from the core of a star to the surface. Sunlight that falls on us today was produced inside the Sun a million years ago.

Neutrinos—"little neutral ones"—are different. They are puzzling, for they have no mass (or so we suspect) and they are electrically neutral. Of most interest is the way they behave. Before SN 1987A, only a handful of neutrinos had ever been detected from space. They appeared to be given off by the Sun and were called solar neutrinos. Inside the Sun, neutrinos are given off rapidly, as photons are. But unlike photons, they move out from the solar core at or near the speed of light, taking only a second or so to exit the core and a few additional seconds to reach the star surface and be thrown into space.

During a violent collapse, when a star reaches a temperature of billions of degrees, a large part of the energy released is in the form of neutrinos. The neutrinos escape quickly, taking energy away from the core of the star in an instant. In a fraction of a second the star cools. The pull of gravity becomes much greater than outward pressure. The star collapses even further. In moments, a large part of the star's mass packs into a space a few miles across. Density becomes so high it is incredible.

As the star collapses and rebounds, its temperature soars once again because of gravitational packing. Any remaining atomic nuclei (nuclear fuel) immediately combine. Elements that are more complex than iron are formed.

The new materials, all of which are in a gaseous state because of the high temperatures, explode and rush outward at thousands of miles a second. For millennia, the gas cloud spreads into space. Eventually, the cloud of gases, together with other clouds that may collide with it, forms into stars.

The enormous explosion of a Type II supernova should release huge numbers of photons. And indeed, SN 1987A did. They were the light that Ian Shelton picked up in his historic photograph of the LMC. Also, there should have been a burst of neutrinos before the photons reached us, and there was. One estimate says that 100 billion neutrinos from SN 1987A passed through you and me, and through every person on our planet. We were not aware of them, because most neutrinos go right through matter; they are not absorbed.

If that is so, how do we know they exist? It's not easy. Huge detectors are needed. Scientists and engineers have built tanks containing thousands of tons of water deep underground beneath tons of solid rock. Most of the neutrinos go right through the Earth and through the water. On very rare occasions, a neutrino reacts with one of the molecules in the liquid. When it does, there is a flash of light that is picked up by one of scores of detectors located around the reservoir. Although neutrinos are believed to abound all the time, days and weeks may go by without a single reaction.

The flow was so much greater just before the time that SN 1987A appeared in the sky that people who read the detectors could not believe what they were seeing. In Japan, the detector, located in a mine at Kamioka, showed eleven events all occurring about a day before the supernova was seen. That's what should happen according to the theory, which says that the neutrinos should come to us directly, while the photons—the light—should be somewhat delayed. Photons are absorbed, you remember, while neutrinos are not.

Another detector is located in the United States,

2,000 feet underground in an abandoned salt mine below Lake Erie. At the same time that the Japanese noted a neutrino "shower," the Americans saw eight reactions in a period of only six seconds. Such an abundance was unheard of. There was no question—SN 1987A was the source of the billions of neutrinos. The high energy level, indicated by the abundance of neutrinos, told scientists that within the interval of a few seconds SN 1987A had produced energy equal to the total output of all the billions of stars in our entire galaxy.

The neutrino abundance, and the timing of their reception a few hours before the light had reached us, was further proof that SN 1987A was not a Type I supernova. In that kind of event, fewer neutrinos, and all at a much lower energy level, are produced.

The explosion also produced tremendous amounts of cosmic ray particles, X rays and gamma rays. Cosmic ray particles are subatomic; they are the cores of atoms. They are ejected from a supernova and speed away in all directions, each carrying tremendous energy. They rain down on planets. Each of us on Earth is bombarded by them.

Laboratory experiments show that cosmic rays and

gamma rays affect genes, the units in our cells that determine our heredity. Genes in reproductive cells determine eye color, body structure, appearance, immunity control, even the way we use our muscles. Largely, it's our genes that make us similar to our parents and grandparents. Genes do change, however. One cause of change is bombardment by high-energy radiation. Very likely, down through the ages, gene alteration by energetic cosmic rays, some of them from supernovas, has caused changes in plants and animals, and has thus very much affected the appearance and functioning of all living things.

Not only are we children of the stars: We are changed by the stars.

6

SANDULEAK −69° 202:

BIRTHPLACE OF
SUPERNOVA 1987A

In many respects, SN 1987A is something of a maverick; it has not behaved exactly in the manner that theory says a Type II supernova should. For example, as you can see from the dashed lines in the diagram on page 43, astronomers expected the exploding star to increase rapidly in brightness. They predicted that after a month or so, brightness would drop off sharply and more or less evenly. After three or four months,

The dashed lines show changes in the brightness of three supernovas. Notice in the beginning there was a large increase in brightness, followed by a steady and even decline.

SN 1987A behaved differently. There was a rapid increase in brightness. Then, most unusually, the energy leveled off for a few days. After that, there was an increase followed by a steady decline.

it should fade away as shown in the dashed lines.

The solid line shows light changes as they actually occurred in SN 1987A. The Supernova increased in brightness to magnitude 4, evened off for a few days, and then brightened to magnitude 3. It be-

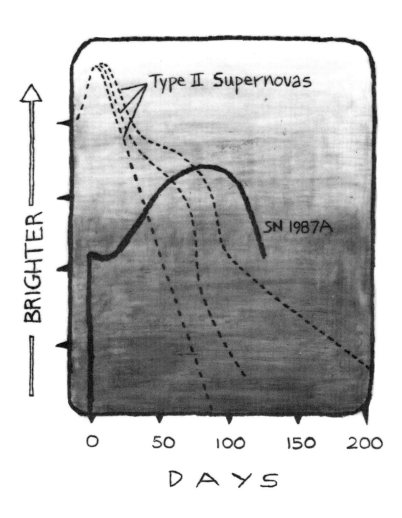

came brighter than the LMC itself, and then faded steadily.

Before SN 1987A, scientists had theorized that Type II supernova explosions always occurred in red supergiant stars. But photographs of the LMC made before the supernova event show no red supergiant in the area. There was, however, a blue supergiant in the region—a star believed to have been 18 times more massive than the Sun. It was called Sanduleak −69° 202 (number 202 in a group of stars located 69 degrees south of the equator). The star was named after Nicholas Sanduleak, an American astronomer who had made a catalog of hundreds of stars that seemed to be of special interest. He was certainly right about this one.

Careful study of the area reveals that the bright-blue star is gone. It seems that Sanduleak −69° 202 must have been the site of the blowup.

Today the exploded star, like an onion, is made of many layers. The layers are in the form of gases thrown out by the explosion, some moving as fast as 25,000 miles a second. Other layers move more slowly.

For much of its eleven-million-year lifetime, Sanduleak −69° 202 was a hydrogen-fusing star, similar to the Sun but much hotter. It was very bright, and so hot it stripped electrons from hydrogen atoms in interstellar space for many light years around. During this time a stellar wind blew bits and pieces of atoms from the star itself out into space. (Right now a similar but much less violent wind is blowing from the Sun.)

After much of the hydrogen of Sanduleak $-69°\,202$ was consumed in fusion reactions and lost through the stellar wind, the star contracted and became even hotter. The intense heating caused the star's outer layers to expand, and it became a red supergiant. According to the theory, one would expect the star to remain a red supergiant while many changes occurred in its core. After less than a million years, the red supergiant would end its life in a supernova explosion. But although those changes did happen in the core, the star did not explode at the red supergiant stage. About 45,000 years before the explosion, the outer layers of the star contracted and the star became a blue supergiant. Then the nuclear fuel in the core was used up. Energy production slowed down, and the star collapsed. The collapse caused the temperature to shoot up once more—this time so high that heavy elements were created. The collapsing material

THE POSSIBLE LIFE HISTORY OF

1 About 11 million years ago the star was converting hydrogen into helium at a rapid rate.

BLUE WHITE

2 RED SUPERGIANT

About 650,000 years before the star exploded, the core contracted and heated. Helium was converted to carbon and oxygen. The outer part of the star was pushed out by the heat; it cooled and turned red. The star became a red supergiant.

3 BLUE SUPERGIANT

45,000 years before the explosion, the outer layers of the star began to contract and get hotter. The star became a blue supergiant. Then the core contracted and heated to 740 million degrees... high enough for carbon to fuse and become neon, sodium, and magnesium.

4 Sixteen years before the explosion, the core temperature reached 1.6 billion degrees. Neon began to fuse to make more oxygen and magnesium.

5 About 4 years before the explosion, oxygen started fusing to make silicon and sulfur.

6 One week before the explosion, silicon and sulfur began changing into iron.

The star was made of layers of different elements ... the lightest at the outside

STAR IN BALANCE

IRON
SILICON
OXYGEN
NEON
CARBON
HELIUM
HYDROGEN

7 MAXIMUM SCRUNCH

As soon as the silicon and sulfur in the star's core had turned to iron, the fusion stopped. Iron could not be fused into heavier elements. In only a few seconds, the core collapsed ...from a sphere about the size of Earth to one about 10 miles across.

of the core rebounded, producing a shock wave that raced outward through the star and blew the star apart. The outer layers, mostly hydrogen and helium, exploded outward at high speed. They gave off ultraviolet light and visible light, which have been detected. Inner parts of the star, containing the heavier elements, were also expelled, although at slower speeds. All that remains of the star, astronomers believe, is this expanding debris and, at the center, a neutron star. If the neutron star really has been detected, it may be rotating faster than any we have known.

7

SUPERNOVA 1987A—
PRESENT AND FUTURE

Interstellar space is not empty; it contains dust and gases. Interstellar gas is made up of mostly hydrogen and some helium. Interstellar dust is made up of the rest of the elements including many kinds of solid particles, such as smokelike grains of carbon and grains of silicate minerals, named after silicon, one of their main ingredients (which, incidentally, is the base ingredient of sand).

As we said earlier, stars form from these clouds of gas and dust as do planets and possibly planetary satellites, comets, and asteroids. Dust and gases are the raw materials out of which all parts of the universe are constructed. And many of these raw materials are produced in supernova explosions.

The heavy elements that are created during the packing, superheating, and exploding of the star's core are blown into space as gas clouds. Once there, some of the gases cool and condense into the fine particles that make the dust clouds.

Light echoes coming from dust clouds 400 and 1,000 light-years from SN 1987A have been detected. Light from the supernova goes out in all directions. Some of the light comes to us directly, and some of the light reaches us only after bouncing off the dust grains in space far outside the star. The light that bounces travels greater distances, and so reaches us later. These later receptions are the so-called light echoes. They enable astronomers to get some idea about how dust clouds are arranged in the region of the supernova.

Also, as we explained earlier, cobalt has been detected in the supernova region. It is clear that a large amount of the supernova material, equal to 25,000

Light echoes, the hazy rings in the photograph, are produced when light from the star bounces off dust clouds located 400 and 1,000 light-years from the supernova itself. (The black spot at the center is caused by a disc used to reduce the glare of the supernova.)

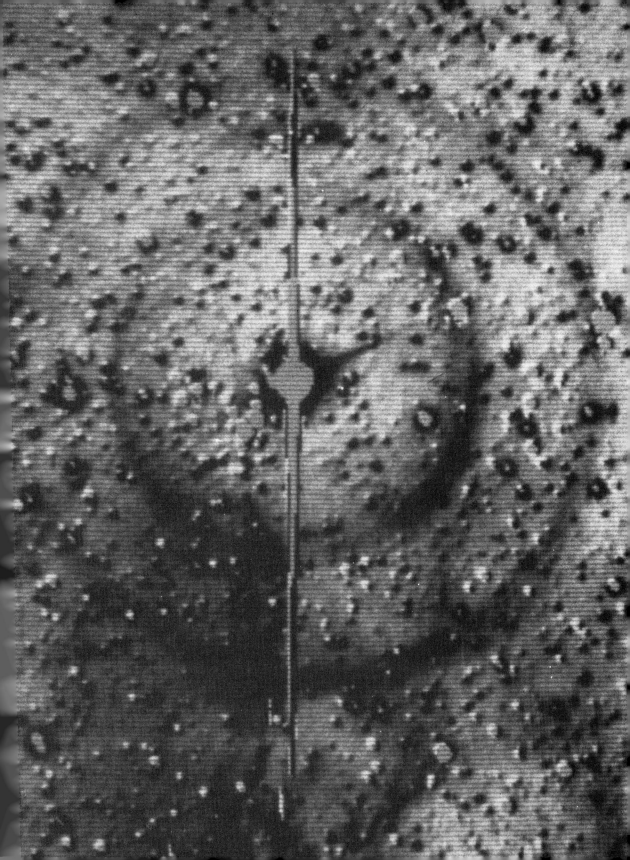

Earths, was converted into radioactive nickel. The element decayed rapidly to become cobalt, which is also radioactive. Cobalt decays into iron. As it does, energy is given off. This energy now keeps the supernova producing visible and infrared radiation.

Over the next few decades, as the outer layers of SN 1987A spread out, astronomers will be able to look further inside. If a pulsar exists there, as some have reported, they will be able to observe it with certainty. This will be the first time that the birth of a pulsar has been watched.

It is possible that the sighting of a pulsar was not accurate. Matter in the star may have collapsed to a density even greater than that of a neutron star. The stellar remnant of SN 1987A may eventually turn out to be a black hole. If so, no energy would be able to escape from it; we would never be able to see it.

Supernova 1987A is the brightest point source ever observed outside our galaxy. And it is the brightest supernova ever observed by modern instruments. It will remain an exciting puzzle for astronomers for a century and more. It took 170,000 years for the energy from SN 1987A to reach us. Right now photons and neutrinos from other supernovas are also speeding toward us. They could arrive at any time. When they do, we can be sure they will tell us even more about how the elements are produced. They will give us further proof that the Earth—and everything on it— is made of stardust.

FURTHER READING

Branley, Franklyn M. *Black Holes, White Dwarfs, and Superstars.* New York: Thomas Y. Crowell, 1976.

———. *Mysteries of the Universe.* New York: E. P. Dutton & Co., 1984.

Kippenhahn, Rudolph. *100 Billion Suns: The Birth, Life, and Death of the Stars.* New York: Basic Books, 1983.

Kirshner, Robert P. "Supernova—Death of a Star." *National Geographic,* May 1988.

Murdin, Paul, and Lesley Murdin. *Supernovae.* New York: Cambridge University Press, 1985.

Stephenson, F. Richard. "Guest Stars Are Always Welcome." *Natural History,* September 1987.

Tierney, John. "Exploding Star Contains Atoms of Elvis Presley's Brain." *Discover,* July 1987.

Woosley, Stanford E., and M. M. Phillips. "Supernova 1987A!" *Science,* May 6, 1988.

INDEX

References to illustrations are in *italic* type.

dust clouds, 9, 49, 50

energy
    as cosmic ray particles, 40
    and light production, 6, 28, 31
    as neutrinos, 37
    of neutron stars, 31, 32
    as radio waves, 6, 31

galaxies
    Large Magellanic Cloud
        (LMC), *vi*, 1–2
gamma rays, 6, 10, 40, 41
gas clouds, *vi*
    and planet formation, 9
    in star formation, 38
    from supernovas, 17, 32, 50
    See also Crab nebula
gases
    interstellar, 49
    and star formation, 12, *13*
    in Type I supernovas, 15, 17
gene alteration
    and high-energy radiation, 41
gravity
    in star collapse, 37–38
    and star equilibrium, 14
    of white dwarf star, 15, 17
guest stars, 22

heavy elements
    as dust cloud particles, 50
    formation of, 7–8, 11, 20, 38
helium, *6*
    fusion of, 6–7, 18
    in interstellar gas, 49

and star formation, 4, 6
Hipparchus, 27
hydrogen, *6*
    fusion of, 4, 6
    in interstellar gas, 49
    as star atmosphere, 20
    and star formation, 4, 6
    in supernovas, 18, 35

International Ultraviolet
    Explorer, 35
interstellar gas, 49
iron, formation of, 7, 10, 20, 52

Kepler, Johannes, 23, *25*, 26

Large Magellanic Cloud (LMC),
    vi, 1–2, 38, 44, *color insert*
Las Campanas Observatory, 1,
    color insert
light
    from nuclear fusion, 6
    as photon, 35–37, 39
    from SN 1987A, 42, 43, 44, 48
    See also brightness
light echoes, 50, *51*

nebula. *See* gas clouds
neutrino detectors, 39–40
neutrinos
    behavior of, *36*, 37, 38, 39
    release by SN 1987A, 38, 39–40,
        color insert
neutron star, *19, 30*
    creation of, 21, 30–31, 32
    from SN 1987A, 48
    See also pulsar

nova (*stella nova*)
    definition of, 2
nuclear fusion reaction
    neutrinos produced by, 37–38
    photons produced by, 35–37
    in star formation, 6, 7
    in Type II supernovas, 18, 20

oxygen
    from helium fusion, 7, 18, 20

Parsons, William, 28
photons
    behavior of, *36*, 36–37, 39
    from nuclear fusion, 35–37
    release by SN 1987A, 38
planet formation, 9, 10, 49
pulsar, *33*
    in Crab nebula, 30
    definition of, 31–32
    in SN 1987A, 52

radio waves
    from Crab nebula, 30, 31
    from nuclear fusion, 6
    in pulsar, 31–32
    in Type I supernovas, 17
red giant star, 7
red supergiant star, 44, 45

Sanduleak, Nicholas, 44
Sanduleak –69° 202, x, *46, 47,*
    color insert
    as blue supergiant, 44, 45
    explosion of, 45, 48
    location and mass of, 44

See also
    Supernova 1987A
Shelton, Ian, 1–2, 38, *color insert*
shock wave, 9
    from Sanduleak –69° 202, 48
    in Type II supernovas, *19,*
        20–21
silicon
    formation of, 7
    fusion of, 20
    in interstellar dust, 49
solar neutrinos, 37
solar system formation, 9
stars
    blue supergiant, 44
    brightness of, 27
    collapse of, 37–38
    in double-star system, 15, 17
    formation of, 4, 6, 9–10, 12, *13,*
        14, 38
    red supergiant, 44–45
    as supernovas, 2–3
    white dwarf, 7, 15, 17
stella nova, 2
stellar wind, 45
Sun, *5*
    age and mass of, 14
    brightness of, 27
    equilibrium in, 14
    formation of, 10
    future of, 6–7
    neutrinos from, 37
    photons from, 37
    temperature of, 18
Supernova 1987A, *color insert*
    black hole in, 52

57